Cone Beam CT of the Head and Neck

Chung How Kau • Kenneth Abramovitch
Sherif Galal Kamel • Marko Bozic

Cone Beam CT of the Head and Neck

An Anatomical Atlas

Springer

Professor Chung How Kau
University of Alabama
Birmingham School of Dentistry
Department of Orthodontics
7th Avenue South 1919
35294 Birmingham Alabama
Room 305, USA
ckau@uab.edu

Dr. Sherif Galal Kamel
University Hospital of Coventry
and Warwickshire
Clifford Bridge Road
CV2 2DX Coventry
Medical Residence Room 1-9A
UK
dr.sherif83@gmail.com

Dr. Kenneth Abramovitch
University of Texas
Health Science Center
Department of Diagnostic Sciences
Section for Oral Radiology
MD Anderson Blvd. 6516
77030 Houston Texas, USA
Kenneth.Abramovitch@uth.tmc.edu

Dr. Marko Bozic
University Medical Center
Dept. of Maxillofacial and Oral
Surgery
Zaloska 2
1525 Ljubljana
Slovenia
marko.bozzich@gmail.com

ISBN 978-3-642-12703-8 ISBN 978-3-642-12704-5 (eBook)
DOI 10.1007/978-3-642-12704-5

Springer Heidelberg Dordrecht London New York

Library of Congress Control Number: 2010932935

Cover design: eStudioCalamar, Figueres/Berlin

Printed on acid-free paper

Springer is part of Springer Science+Business Media (www.springer.com)

Acknowledgments

The following people have been involved in the project without which this atlas would never have been completed:

1. ProfessorM arkW ong
2. Dr.Fe nP an
3. Dr.H asmatPopa t
4. MissJ enniferN guyen
5. Mr.K urtE C lark
6. Dr.N adaSouc car

Contents

Introduction

C.H. Kau et al., *Cone Beam CT of the Head and Neck*,
DOI: 10.1007/978-3-642-12704-5_1, © Springer-Verlag Berlin Heidelberg 2011

This pocket clinical atlas was produced to help dental and medical colleagues understand and correlate structures of the head and neck with cone beam computerized tomography (CBCT) imaging technology.

Methods used for radiographic evaluation and diagnosis have undergone enormous changes in the last 20 years. New technologies are being developed and are becoming readily available to the medical and dental field. The advancements in hardware and software have allowed the development of innovative methods for facial diagnosis, treatment planning, and clinical application.

CBCT was developed in the 1990s as an advancement in technology resulting from the demand for three-dimensional (3D) information obtained by conventional computed tomography (CT) scans. The development of CBCT technology reduces exposure by using lower radiation dose, compared with conventional CT [1–3]. As the demand for the technology increases, so has the market for custom built cranio-maxillofacial CBCT devices. The rates of increase for CBCTs have been increasing in number on the market over the last decade and a variety of applications to the facial and dental environments have been established[2].

ConventionalC omputedT omography

CT technology was developed by Sir Godfrey Hounsfield in 1967 and there has been a gradual evolution to five generations of the system. First generation scanners consisted of a single radiation source and a single detector and information was obtained slice by slice. The second generation was introduced as an improvement and multiple detectors were incorporated within the plane of the scan. The third generation was made possible by the advancement in detector and data acquisition technology. These large detectors reduced the need for the beam to translate around the object to be measured and were often known as the "fan-beam" CTs. Ring artifacts were often seen on the images captured distorting the 3D image and obscuring certain anatomical landmarks. The fourth generation was developed to counter this problem. A moving radiation source and a fixed detector ring were introduced. This meant that modifications to the angle of the radiation source had to be taken into account and more scattered radiation was seen. Finally the fifth (sometimes known as the sixth) generation scanners were the introduction to reduced "motion" or "scatter" artifacts. As with the previous two generations, the detector is stationary and the electron beam is electronically swept along a semicircular tungsten strip anode. Projections of the X-rays are so rapid that even the heart beat may be captured. This has led some clinicians to hail it as a 4D motion capture device [2, 4]. In 2007, the Toshiba "dynamic volume" scanner based on 320 slices is showing the potential to significantly reduce radiation exposure by eliminating the requirement for a helical examination in both cardiac CT angiography and whole brain perfusion studies for the evaluation of stroke [online reference, 1].

There are, however, limitations to these CT systems. They are expensive and require a lot of space. The 3D reconstruction is time consuming and so less cost efficient. Furthermore the radiation exposure to the patient has limited their usage to complex craniofacial problems and for specialized diagnostic information only.

CBCT

CBCTs for dental, oral, and maxillofacial surgery and orthodontic indications were designed to counter some of the limitations of the conventional CT scanning devices. The radiation source consists of a conventional low-radiation X-ray tube and the resultant beam is projected onto a flat panel detector (FPD) or a charge-coupled device (CCD) with an image intensifier. The FPD was shown to have a high spatial resolution [5]. The cone beam produces a more focused beam and much less radiation scatter compared with the conventional fan-shaped CT devices [6]. This significantly increases the X-ray utilization and reduces the X-ray tube capacity required for volumetric scanning [7]. It has been reported that the total radiation is approximately 20% of conventional CTs and equivalent to a full mouth periapical radiographic exposure [8]. CBCT can therefore be recommended as a dose-sparing technique compared with alternative standard medical CT scans for common oral and maxillofacial radiographic imaging tasks [9]. The images are comparable to the conventional CTs and may be displayed as a full head view, as a skull view, or as localized regional views.

CBCT Data

The tube and the detector perform one rotation (180 or 360°) around the selected region. The resulting primary data are converted into slice data. The reconstructed slice data can be viewed in user-defined planes. The CT volume consists of a 3D array of image elements, called voxels. Each voxel is characterized with a height, width, and depth. Since the voxel sizes are known from the acquisition, correct measurements can be performed on the images. The spatial resolution in a CT image depends on a number of factors during acquisition (e.g., focal spot, size detector element…) and reconstruction (reconstruction kernel, interpolation process, voxel size). Image noise depends on the total exposure and the reconstruction noise. Increasing the current in the X-ray tube increases the signal-to-noise ratio, and thus reduces the quantum noise of the statistical nature of X-rays, at the expense of patient dose. The artifacts of CT imaging are the consequence of beam hardening, photon scattering, nonlinear partial volume effect, motion, stair step artifact, and others.

Most machines support the digital imaging and communications in medicine (DICOM) format export. The images can therefore be used for most if not all the (software) applications utilized by conventional CT[10].

The following is a summary of the additions and modifications of CBCT as compared with conventional CT, which make CBCT a more appealing alternative:

- Radiation dose is lower. This is mainly because of the lower effective tube current used for the CBCT: while the voltage of the source is approximately the same (90–120 kV), the current is roughly between 1 and 8 mA for the CBCT while e.g., for the multislice CT the current is around 80 mA but can also be as high as 200 mA.
- Detection systems are different (FPD or CCD with image intensifier.
- The resolution is higher; this is mainly due to lesser isotropic voxel size[11].
- There is less artifacts caused by metallic structures but because of lower dose there is more noise and detailed information about soft tissues islos t[10].
- CBCTis le sse xpensivea nds maller.

CBCTA cquisitionSys tems

In 2005, four main CBCT devices were reported in the literature and it was expected that many companies were to enter the market [2]. In July 2008, there were 16 manufacturers of CBCT devices producing 23 different models. There are various classifications of devices but CBCT devices may be divided to fit into four subcategories based on the need of the clinician: and one field of view of the scan (FOV)

- Dentoalveolar (FOV less than 8 cm)
- Maxillo-mandibular (FOV between 8 and 15 cm)
- Skeletal (FOV between 15 and 21 cm)
- Head and neck (FOV above 21 cm)

The important differences besides the clinical classification are the radiation dose, size and weight, time needed for the reconstruction, voxel size, scanning time etc. Furthermore the differences in prices, software, and warranty are important considerations.

Uses of CBCT Technology

Radiographic evaluation and diagnosis have undergone enormous changes in the last 20 years. The important differences between the devices are their FOV, the irradiation dose, size and weight, time needed for the reconstruction, voxel size, scanning time, price, software and warranty. The use of CBCT has many applications. The indications for the CBCT imaging has not been clearly established yet. However, CBCT imaging may be used for the following reasons:

1. Diagnosis
2. Clinicala pplication
3. Clinicale valuationoftre atmentoutc omes

Diagnosis

Common uses of CBCT technology in the head and neck is for impacted teeth evaluation [13], implant treatment planning [14], evaluations of the temporomandibular joint (TMJ) [15], simulations for orthodontic and surgical planning, diagnosis of dento-alveolar pathology, evaluation of the nasal/paranasal sinuses, and pharyngeal airways [16]. Furthermore, craniofacial anomalies, for example cleft patients and those undergoing combined orthodontic and maxillofacial therapy, benefit greatly from CBCT imaging as the technology provides more information than conventional images [17]. There has been a debate on the routine use of technology in orthodontics and further studies are needed [18]. It has also been proven that CBCT is accurate to identify apical periodontitis [19]. A recently suggested CBCT-aided method for determination of root curvature radius allows a more reliable and predictable endodontic planning, which reflects directly on a more efficacious preparation of curved root canals [20]. CBCT provides better diagnostic and quantitative information on periodontal bone levels in three dimensions than conventional radiography [21]. CBCT can also be used for maxillofacial growth and development assessment and dentala gee stimation[22].

Clinical Applications of the CBCT

CBCT provides information for 3D models made by rapid prototyping. The obtained 3D models can serve as a matrix that enables precise planning of operations such as for mini-implant positions in anatomically complex sites [23]. A recent study that included phantoms and human cadavers showed that intraoperative CBCT quantifiably improved surgical performance in all excision tasks and significantly increased surgical confidence. Such intraoperative imaging in combination with real-time tracking and navigation should be of great benefit in delicate procedures in which excision must be executed in close proximity to critical structures [24]. Another study included 179 patients undergoing facial surgery and intraoperative CBCT was used. The acquisition of the data sets was uncomplicated, and image quality was sufficient to assess the postoperative result in all cases [25].

Late Evaluation with CBCT

CBCT is also a tool for the evaluation of surgical and orthodontic treatment. There have not been a lot of papers published but they are increasing in their number as the CBCT is becoming more readily available. CBCT was successfully used to compare the anteroposterior positions of the cleft-side piriform margin and alar base with those of the noncleft side in 52 postoperative unilateral cleft lip patients with no alveolar bone graft [26]. CBCT can be used in combination with 3D soft tissue data obtained with stereo photogrammetry, structured light systems and laser acquisition systems for diagnostic, treatment planning and posttreatment evaluation purposes [27]. Evaluation of the nasal and paranasal sinuses and of the pharyngeal airway is also becomingmo rere levant.

The Purpose of the Clinical Atlas

The purpose of this clinical atlas is to provide every dental or medical colleague with the platform to observe and understand images from the Cone Beam Technology. Great care has been put into the dissection of the cadavers and the reproductions of the slice sections both on the human specimens and the CBCT image.

It is hoped that this will serve as a reference for all who are working in the area of CBCT imaging of the head and neck.

References

1. Tam, K.C., S. Samarasekera, and F. Sauer, Exact cone beam CT with a spiral scan. Phys Med Biol, 1998. 43(4): p. 1015-24.
2. Kau, C.H., et al., Three-dimensional cone beam computerized tomography in orthodontics. J Orthod, 2005. 32(4): p. 282-93.
3. Tsiklakis, K., et al., Dose reduction in maxillofacial imaging using low dose Cone Beam CT. Eur J Radiol, 2005. 56(3): p. 413-7.
4. Robb, R.A., X-ray computed tomography: an engineering synthesis of multiscientific principles. Crit Rev Biomed Eng, 1982. 7(4): p. 265-333.
5. Baba, R., et al., Comparison of flat-panel detector and image-intensifier detector for cone-beam CT. Comput Med Imaging Graph, 2002. 26(3): p. 153-8.
6. Mah, J. and D. Hatcher, Current status and future needs in craniofacial imaging. Orthod Craniofac Res, 2003. 6(Suppl 1): p. 10-6; discussion 179-82.
7. Sukovic, P., Cone beam computed tomography in craniofacial imaging. Orthod Craniofac Res, 2003. 6 Suppl 1: p. 31-6; discussion 179-82.
8. Mah, J.K., et al., Radiation absorbed in maxillofacial imaging with a new dental computed tomography device. Oral Surg Oral Med Oral Pathol Oral Radiol Endod, 2003. 96(4): p. 508-13.
9. Ludlow, J.B. and M. Ivanovic, Comparative dosimetry of dental CBCT devices and 64-slice CT for oral and maxillofacial radiology. Oral Surg Oral Med Oral Pathol Oral Radiol Endod, 2008. 106(1): p. 930-8.
10. Swennen, G.R.J., F. Schutyser, and J.-E. Hausamen, Three-dimensional cephalometry: a color atlas and manual. 1st. ed. 2006, Berlin: Springer. xxi, 365 p.
11. Hashimoto, K., et al., A comparison of a new limited cone beam computed tomography machine for dental use with a multidetector row helical CT machine. Oral Surg Oral Med Oral Pathol Oral Radiol Endod, 2003. 95(3): p. 371-7.
12. The 2007 Recommendations of the International Commission on Radiological Protection. ICRP publication 103. Ann ICRP, 2007. 37(2-4): p. 1-332.

13. Nakajima, A., et al., Two- and three-dimensional orthodontic imaging using limited cone beam-computed tomography. Angle Orthod, 2005. 75(6): p. 895-903.
14. Madrigal, C., et al., Study of available bone for interforaminal implant treatment using cone-beam computed tomography. Med Oral Patol Oral Cir Bucal, 2008. 13(5): p. E307-12.
15. Honda, K., et al., Osseous abnormalities of the mandibular condyle: diagnostic reliability of cone beam computed tomography compared with helical computed tomography based on an autopsy material. Dentomaxillofac Radiol, 2006. 35(3): p. 152-7.
16. Maki, K., et al., Computer-assisted simulations in orthodontic diagnosis and the application of a new cone beam X-ray computed tomography. Orthod Craniofac Res, 2003. 6(Suppl 1): p. 95-101; discussion 179-82.
17. Korbmacher, H., et al., Value of two cone-beam computed tomography systems from an orthodontic point of view. J Orofac Orthop, 2007. 68(4): p. 278-89.
18. Silva, M.A., et al., Cone-beam computed tomography for routine orthodontic treatment planning: a radiation dose evaluation. Am J Orthod Dentofacial Orthop, 2008. 133(5): p. 640 e1-5.
19. Estrela, C., et al., Accuracy of cone beam computed tomography and panoramic and periapical radiography for detection of apical periodontitis. J Endod, 2008. 34(3): p. 273-9.
20. Estrela, C., et al., Method for determination of root curvature radius using cone-beam computed tomography images. Braz Dent J, 2008. 19(2): p. 114-8.
21. Mol, A. and A. Balasundaram, In vitro cone beam computed tomography imaging of periodontal bone. Dentomaxillofac Radiol, 2008. 37(6): p. 319-24.
22. Yang, F., R. Jacobs, and G. Willems, Dental age estimation through volume matching of teeth imaged by cone-beam CT. Forensic Sci Int, 2006. 159(Suppl 1): p. S78-83.
23. Kim, S.H., et al., Surgical positioning of orthodontic mini-implants with guides fabricated on models replicated with cone-beam computed tomography. Am J Orthod Dentofacial Orthop, 2007. 131(4 Suppl): p. S82-9.
24. Chan, Y., et al., Cone-beam computed tomography on a mobile C-arm: novel intraoperative imaging technology for guidance of head and neck surgery. J Otolaryngol Head Neck Surg, 2008. 37(1): p. 81-90.
25. Pohlenz, P., et al., Clinical indications and perspectives for intraoperative cone-beam computed tomography in oral and maxillofacial surgery. Oral Surg Oral Med Oral Pathol Oral Radiol Endod, 2007. 103(3): p. 412-7.
26. Miyamoto, J., et al., Evaluation of cleft lip bony depression of piriform margin and nasal deformity with cone beam computed tomography: "retruded-like" appearance and anteroposterior position of the alar base. Plast Reconstr Surg, 2007. 120(6): p. 1612-20.
27. Lane, C. and W. Harrell, Jr., Completing the 3-dimensional picture. Am J Orthod Dentofacial Orthop, 2008. 133(4): p. 612-20.

Axial

C.H. Kau et al., *Cone Beam CT of the Head and Neck*,
DOI: 10.1007/978-3-642-12704-5_2, © Springer-Verlag Berlin Heidelberg 2011

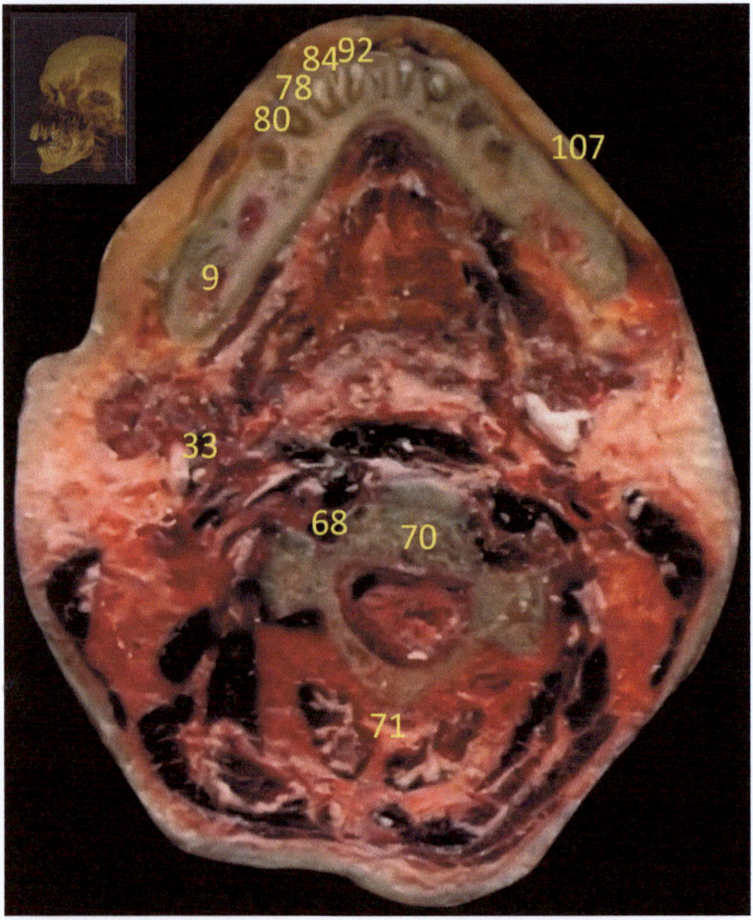

9 – Mandible-body	78 – Mandibular cuspid (Root)
33 – Temporal bone-styloid process	80 – Mandibular first bi-cuspid (Root)
68 – C2 Axis-transverse process	84 – Mandibular lateral incisor teeth (Root)
70 – C2 Axis-body	92 – Maxillary central incisor teeth (Root)
71 – C2 Axis-spine	107 – Face

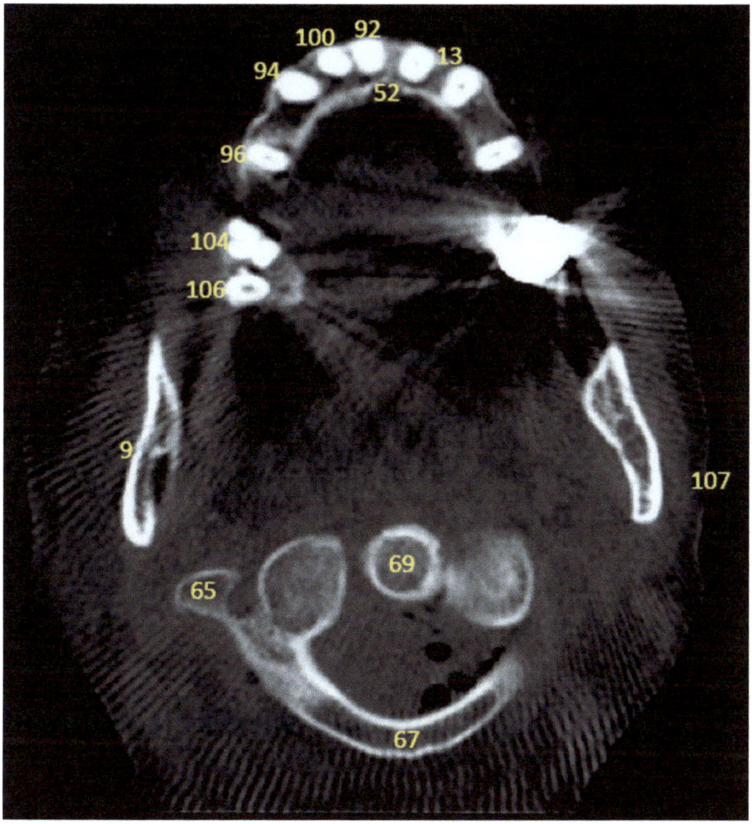

9 – Mandible-body
13 – Maxilla-alveolar bone
52 – Incisive canal
65 – C1 Atlas-body
67 – C1 Atlas-posterior arch/tubercle
69 – C2 Axis-dens
92 – Maxillary central incisor teeth (Root)
94 – Maxillary cuspid (Root)
96 – Maxillary first bi-cuspid (Root)
100 – Maxillary lateral incisor teeth (Root)
104 – Maxillary second molar (Root)
106 – Maxillary third molar (Root)
107 – Face

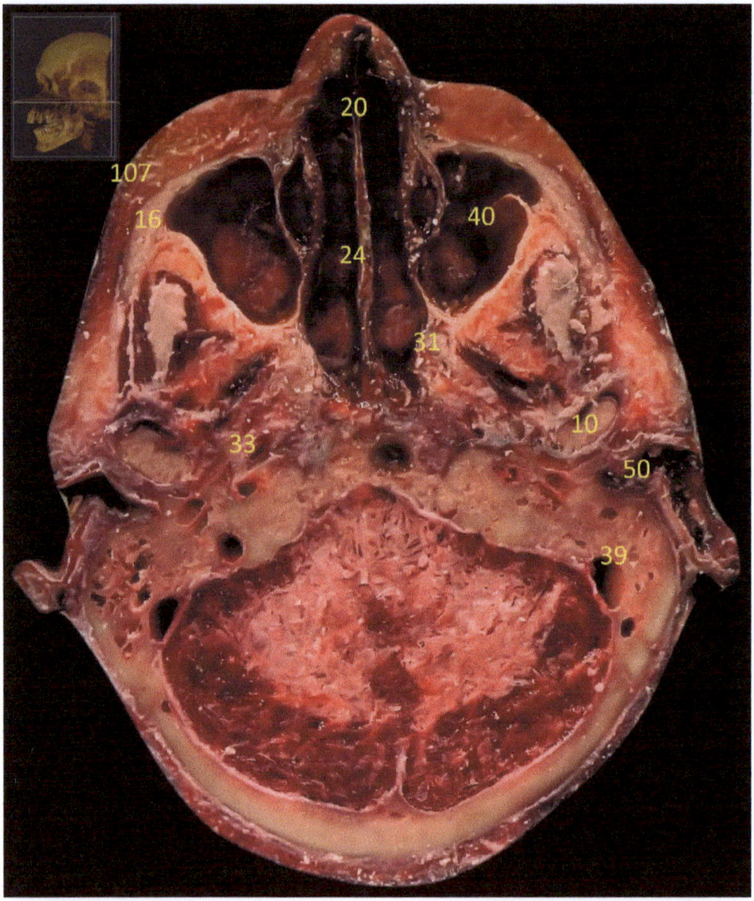

10 – Mandible-condyle	33 – Temporal bone-styloid process
16 – Maxilla-zygomatic process	39 – Mastoid air cells
20 – Nasal septum	40 – Maxillary sinus
24 – Palatine bone-*vomer*	50 – Auricular canal
31 – Sphenoid bone	107 – Face

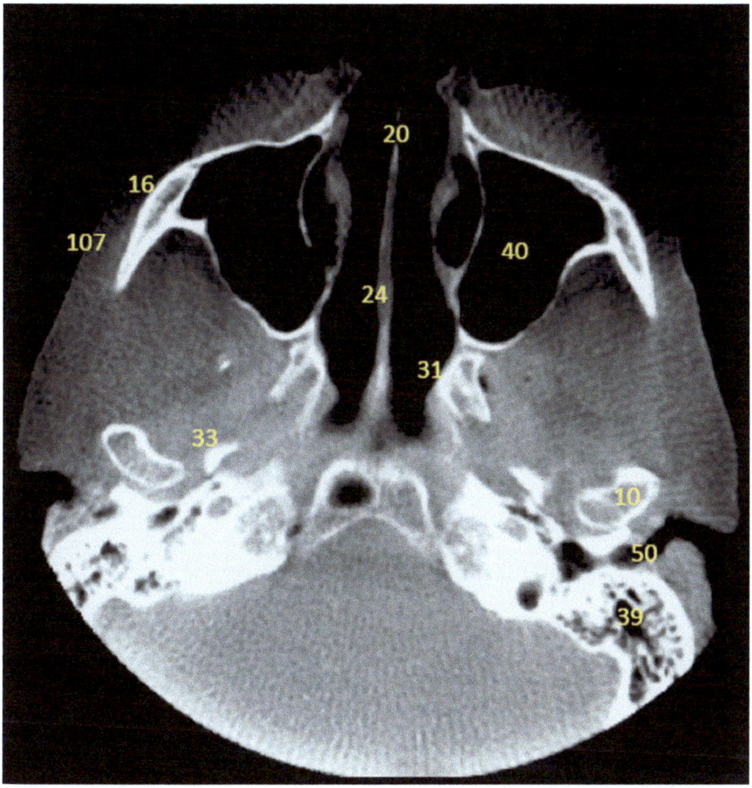

10 – Mandible-condyle
16 – Maxilla-zygomatic process
20 – Nasal septum
24 – Palatine bone – *vomer*
31 – Sphenoid bone
33 – Temporal bone-styloid process
39 – Mastoid air cells
40 – Maxillary sinus
50 – Auricular canal
107 – Face

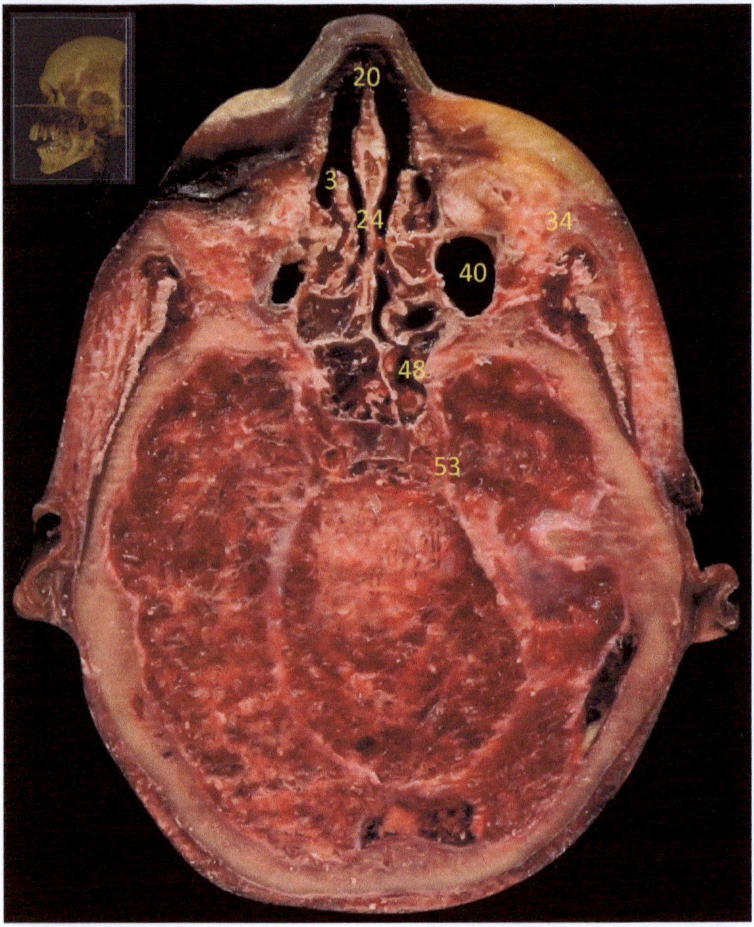

3 – Concha-superior
20 – Nasal septum
24 – Palatine bone-*vomer*
34 – Zygoma
40 – Maxillary sinus
48 – Sphenoid sinus
53 – Internal carotid artery

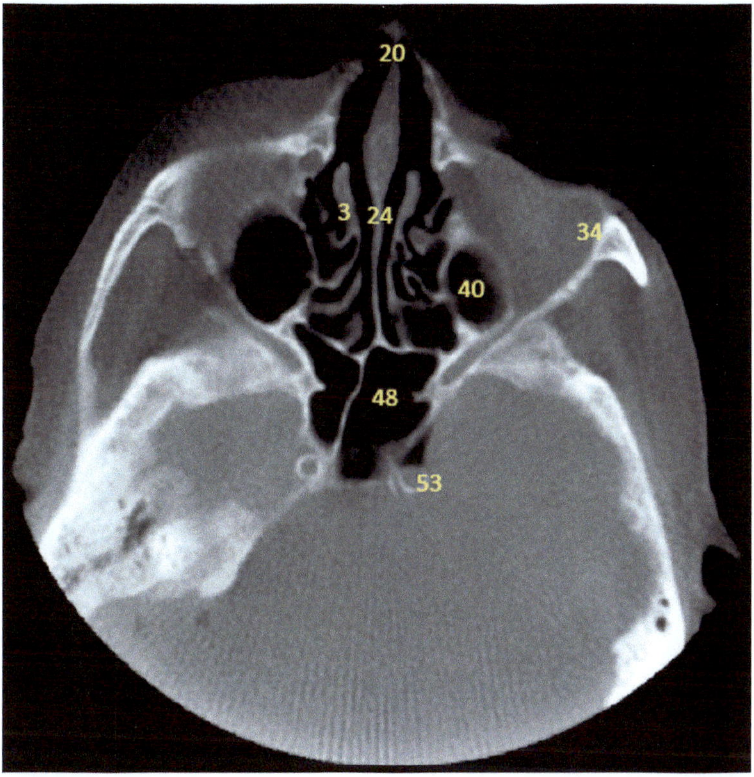

3 – Concha-superior
20 – Nasal septum
24 – Palatine bone-*vomer*
34 – Zygoma
40 – Maxillary sinus
48 – Sphenoid sinus
53 – Internal carotid artery

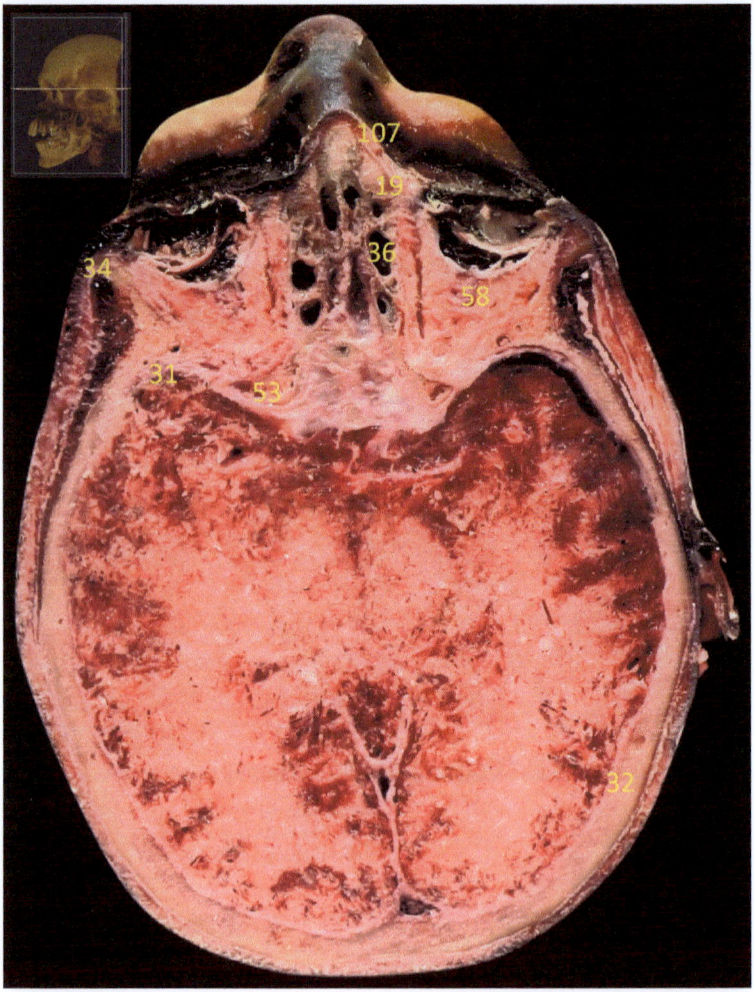

19 – Nasal bone	36 – Ethmoid sinus
31 – Sphenoid bone	53 – Internal carotid artery
32 – Temporal bone	58 – Orbit
34 – Zygoma	107 – Face

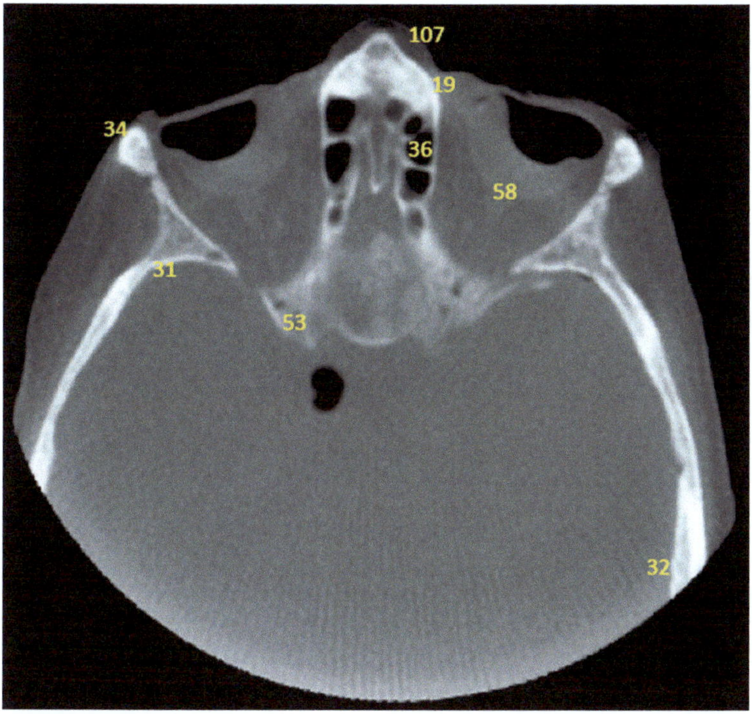

19 – Nasal bone
31 – Sphenoid bone
32 – Temporal bone
34 – Zygoma
36 – Ethmoid sinus
53 – Internal carotid artery
58 – Orbit
107 – Face

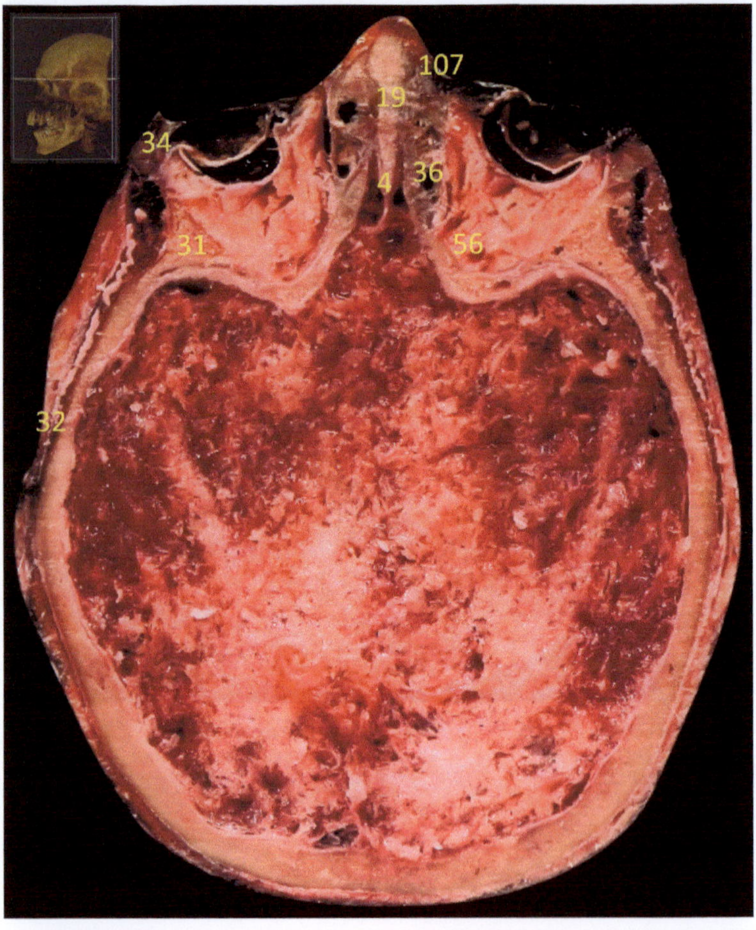

4 – Ethmoid bone-*crista galli*
19 – Nasal bone
31 – Sphenoid bone
32 – Temporal bone
34 – Zygoma
36 – Ethmoid sinus
56 – Optic nerve canals
107 – Face

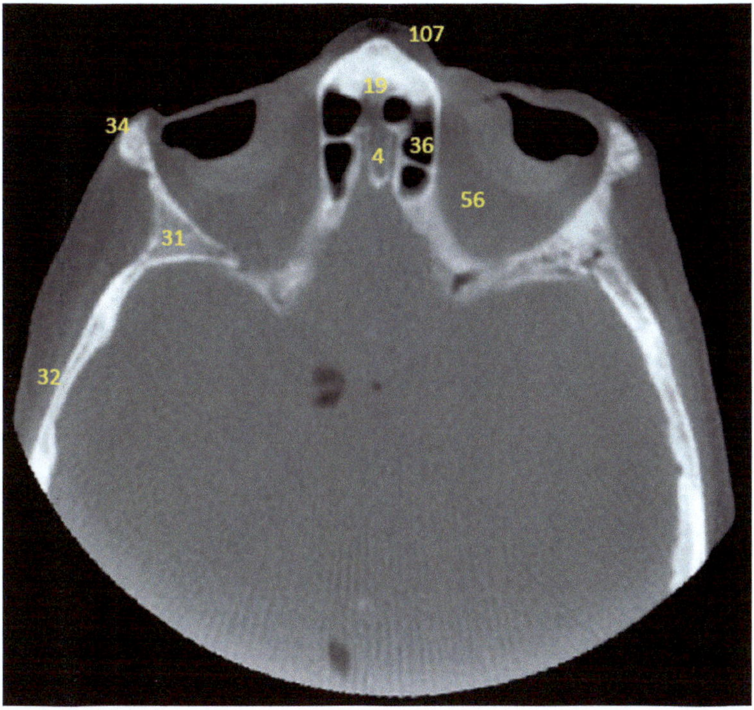

4 – Ethmoid bone-*Crista galli*
19 – Nasal bone
31 – Sphenoid bone
32 – Temporal bone
34 – Zygoma
36 – Ethmoid sinus
56 – Optic nerve canals
107 – Face

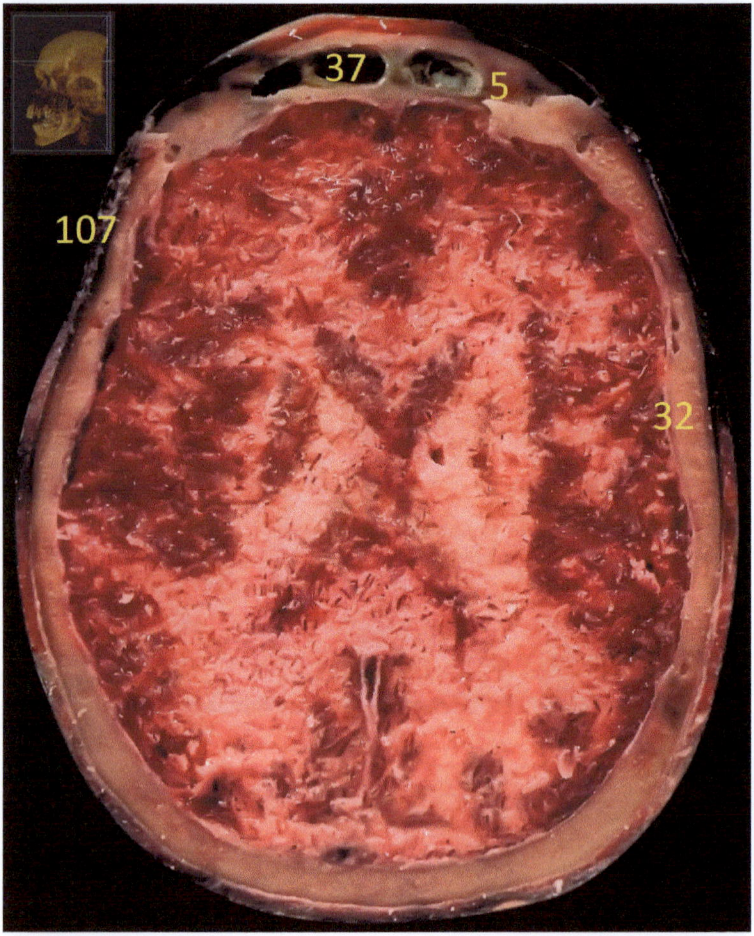

5 – Frontal bone
32 – Temporal bone
37 – Frontal sinus
107 – Face

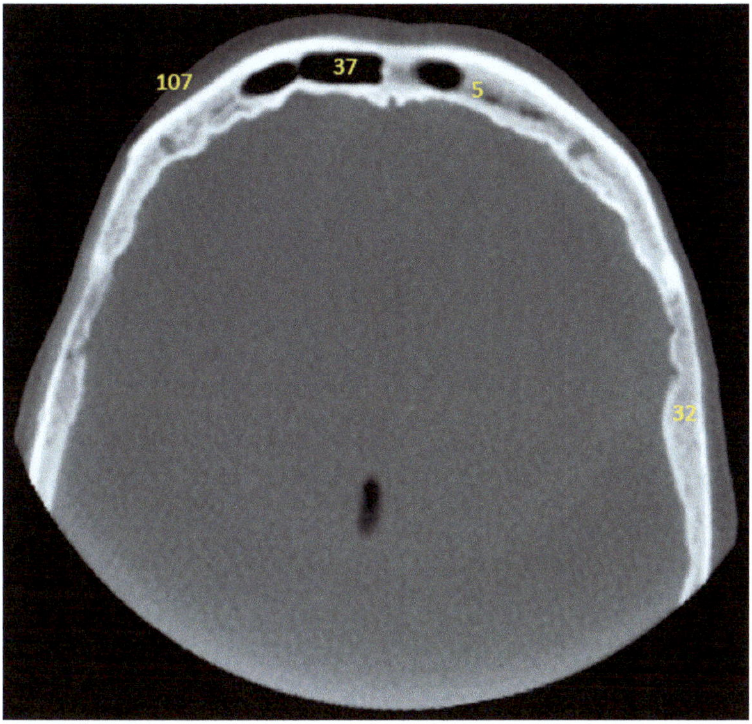

5 – Frontal bone
32 – Temporal bone
37 – Frontal sinus
107 – Face

Coronal

C.H. Kau et al., *Cone Beam CT of the Head and Neck*,
DOI: 10.1007/978-3-642-12704-5_3, © Springer-Verlag Berlin Heidelberg 2011

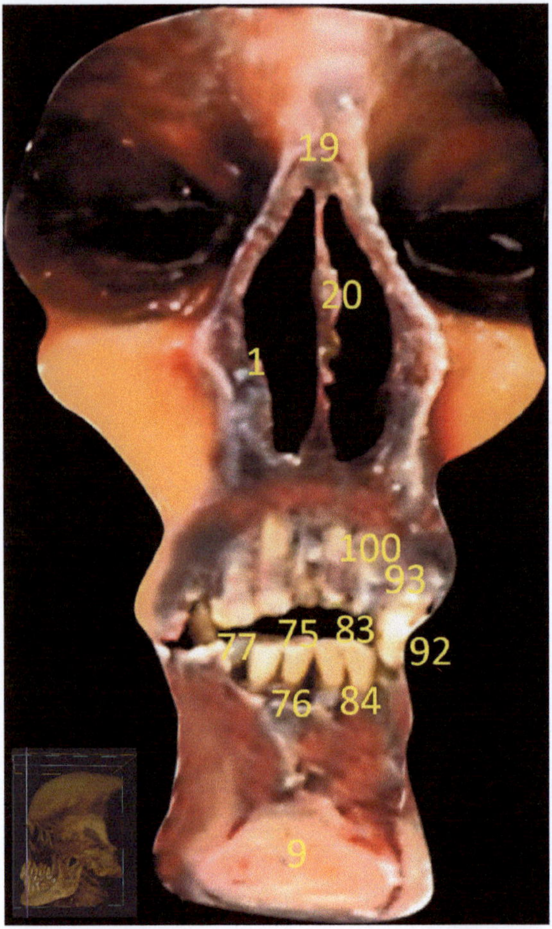

1 – Concha-inferior	77 – Mandibular cuspid (Crown)
9 – Mandible-body	83 – Mandibular lateral incisor teeth (Crown)
19 – Nasal bone	84 – Mandibular lateral incisor teeth (Root)
20 – Nasal septum	94 – Maxillary cuspid teeth (Root)
75 – Mandibular central incisor teeth (Crown)	93 – Maxillary cuspid (Crown)
76 – Mandibular central incisor teeth (Root)	100 – Maxillary lateral incisor teeth (Root)

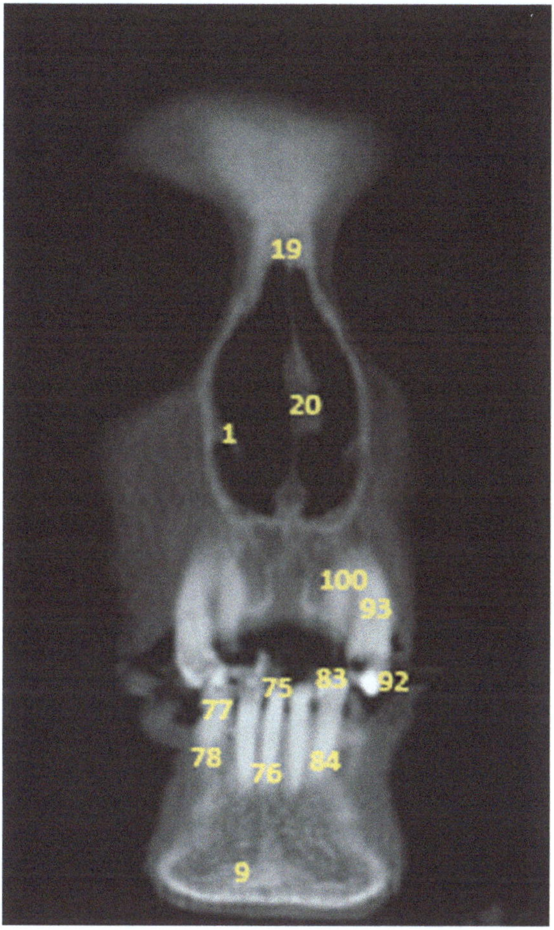

1 – Concha-inferior
9 – Mandible-body
19 – Nasal bone
20 – Nasal septum
75 – Mandibular central incisor teeth (Crown)
76 – Mandibular central incisor teeth (Root)
77 – Mandibular cuspid (Crown)
83 – Mandibular lateral incisor teeth (Crown)
84 – Mandibular lateral incisor teeth (Root)
94 – Maxillary cuspid teeth (Root)
93 – Maxillary cuspid (Crown)
100 – Maxillary lateral Incisor teeth (Root)

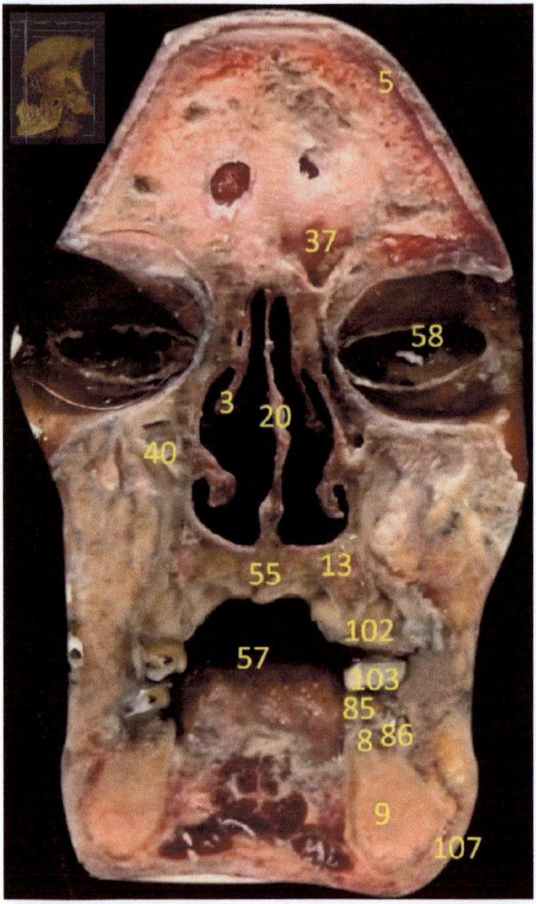

3 – Concha-superior
5 – Frontal bone
8 – Mandible-alveolar bone
9 – Mandible-body
13 – Maxilla-alveolar bone
20 – Nasal septum
37 – Frontal sinus
40 – Maxillary sinus
55 – Mid palatal suture
57 – Oral cavity
58 – Orbit
85 – Mandibular second bi-cuspid (Crown)
86 – Mandibular second bi-cuspid (Root)
102 – Maxillary second bi-cuspid (Root)
103 – Maxillary second molar (Crown)
107 – Face

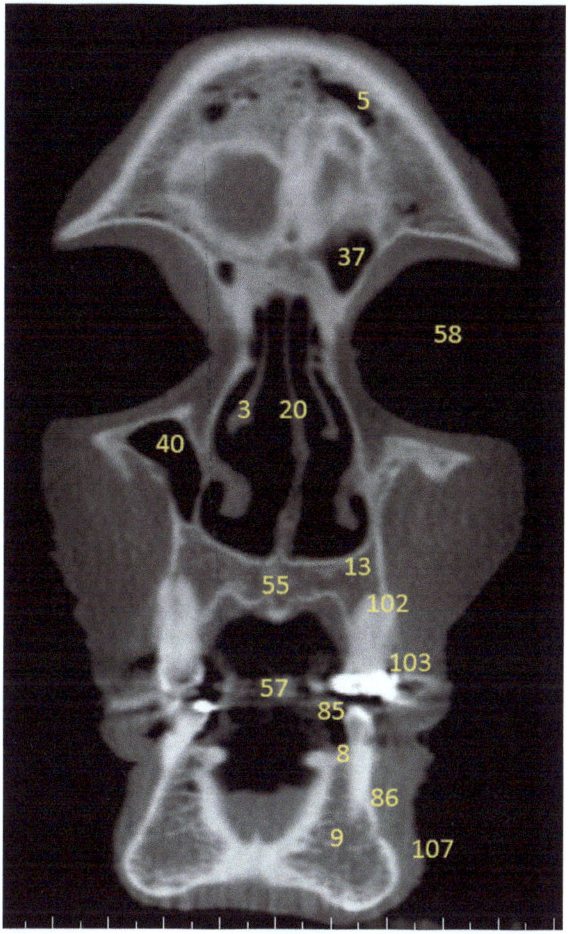

3 – Concha-superior	55 – Mid palatal suture
5 – Frontal bone	57 – Oral cavity
8 – Mandible-alveolar bone	58 – Orbit
9 – Mandible-body	85 – Mandibular second bi-cuspid (Crown)
13 – Maxilla-alveolar bone	86 – Mandibular second bi-cuspid (Root)
20 – Nasal septum	102 – Maxillary second bi-cuspid (Root)
37 – Frontal sinus	103 – Maxillary second molar (Crown)
40 – Maxillary sinus	107 – Face

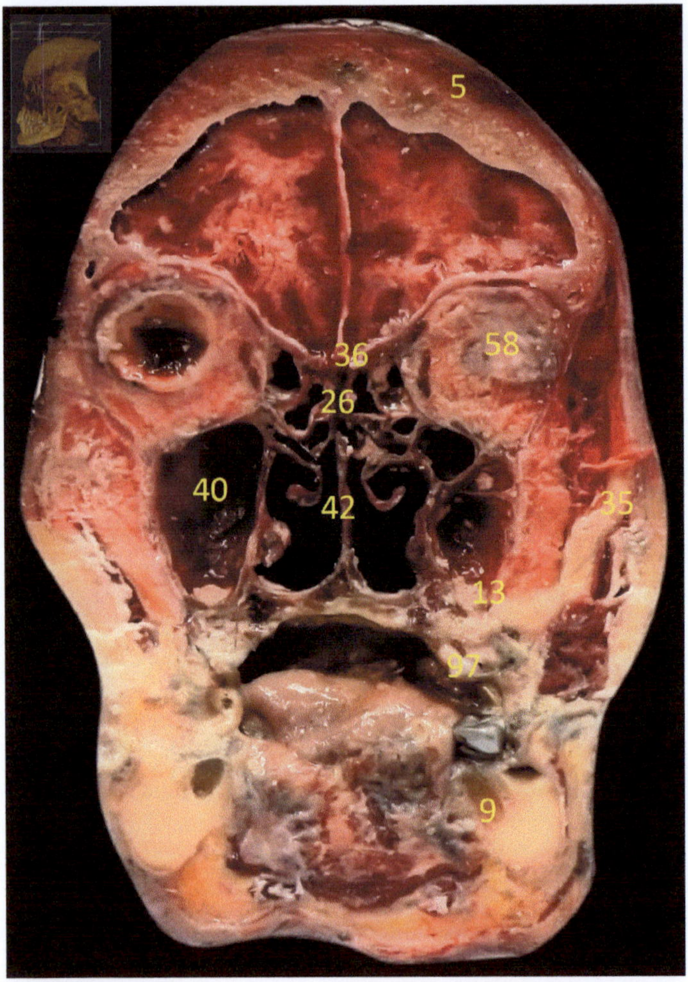

5 – Frontal bone
9 – Mandible-body
13 – Maxilla-alveolar bone
26 – Perpendicular plate of the ethmoid sinus
35 – Zygomatic arch
36 – Ethmoid sinus
40 – Maxillary sinus
42 – Nasal septum
58 – Orbit
96 – Maxillary first bi-cuspid (Root)
97 – Maxillary first molar (Crown)

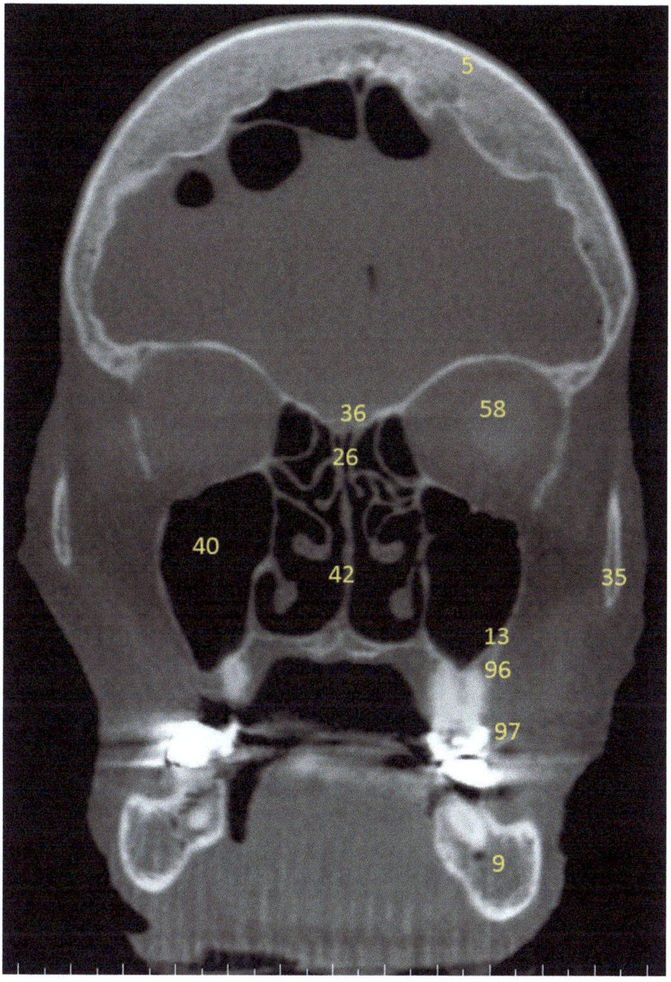

5 – Frontal bone	36 – Ethmoid sinus
9 – Mandible-body	40 – Maxillary sinus
13 – Maxilla-alveolar bone	42 – Nasal septum
26 – Perpendicular plate of the ethmoid sinus	58 – Orbit
35 – Zygomatic arch	96 – Maxillary first bi-cuspid (Root)
	97 – Maxillary first molar (Crown)

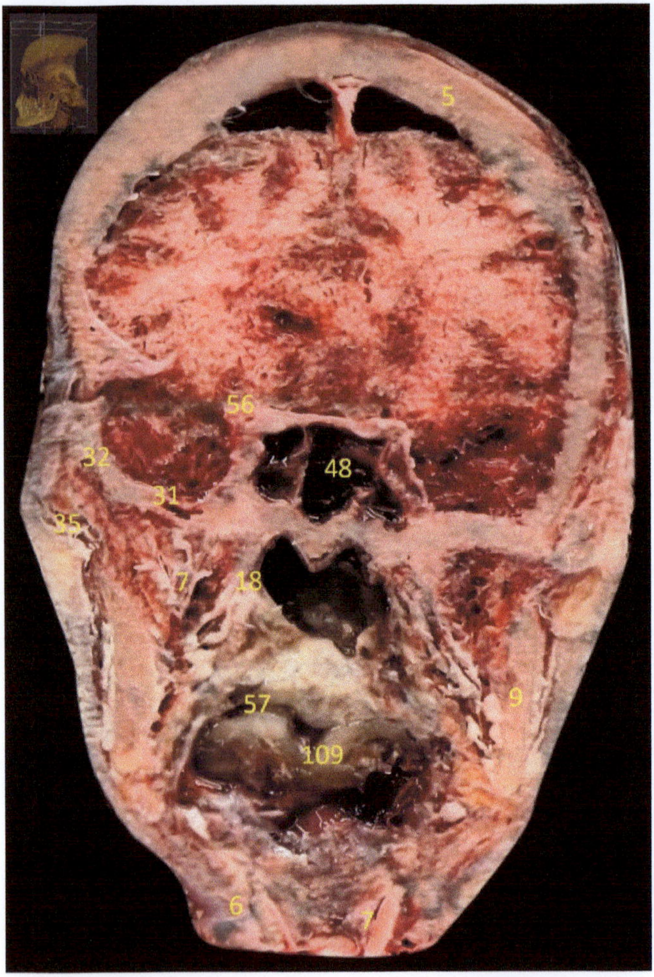

5 – Frontal bone
6 – Hyoid
7 – Lateral pterygoid plate
9 – Mandible-body
18 – Medial pterygoid plate
31 – Sphenoid bone
32 – Temporal bone
35 – Zygomatic arch
48 – Sphenoid sinus
56 – Optic nerve canals
57 – Oral cavity
109 – Tongue

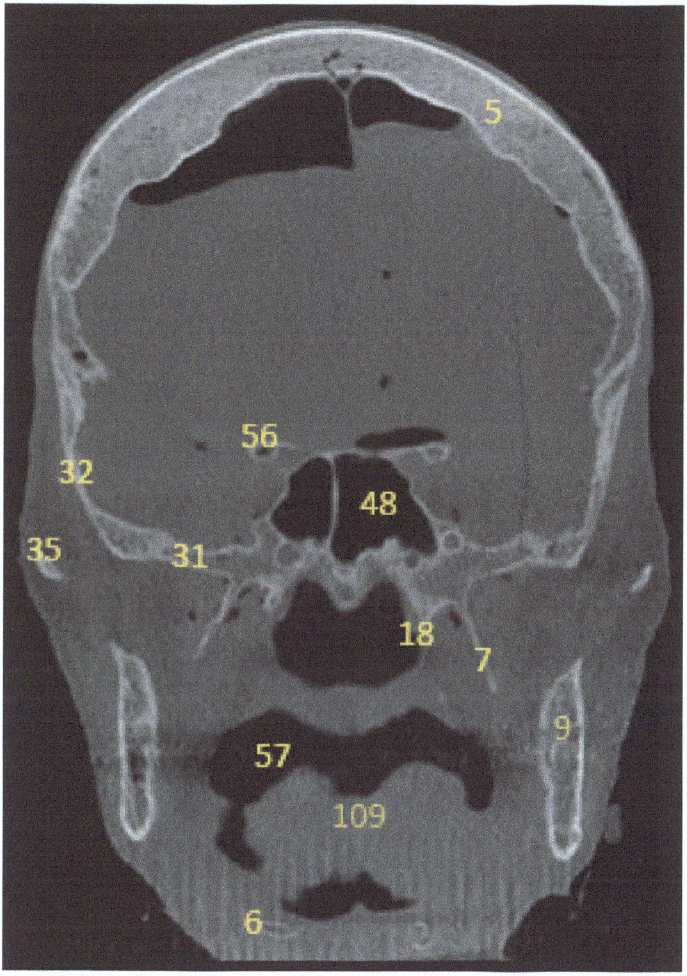

5 – Frontal bone
6 – Hyoid
7 – Lateral pterygoid plate
9 – Mandible-body
18 – Medial pterygoid plate
31 – Sphenoid bone
32 – Temporal bone
35 – Zygomatic arch
48 – Sphenoid sinus
56 – Optic nerve canals
57 – Oral cavity
109 – Tongue

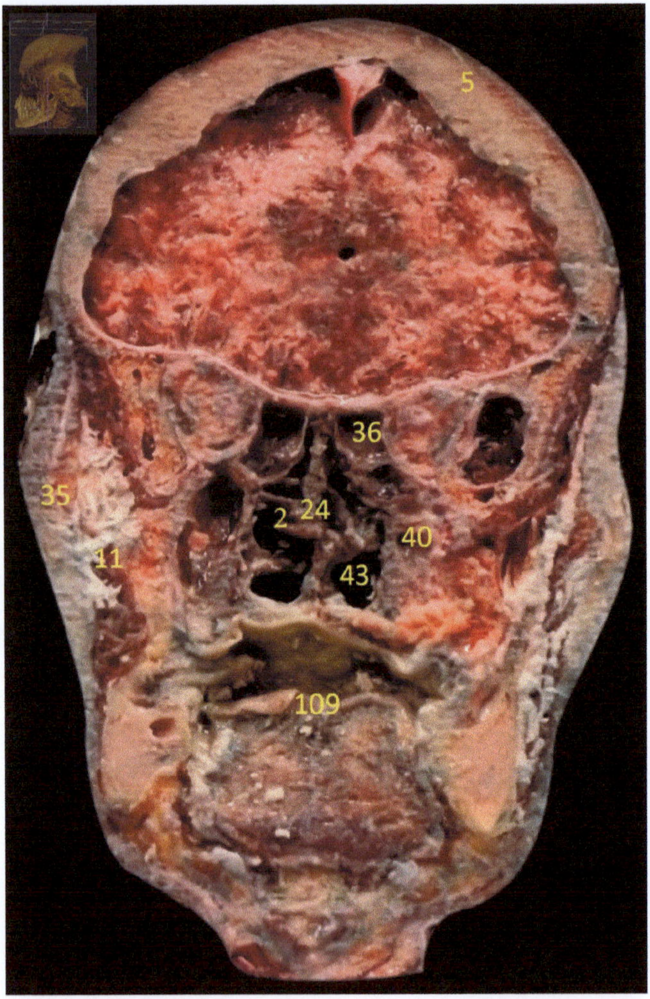

2 – Concha-middle
5 – Frontal bone
11 – Mandible-coronoid process
24 – Palatine bone-*vomer*
35 – Zygomatic arch
36 – Ethmoid sinus
40 – Maxillary sinus
43 – Nasal sinus
109 – Tongue

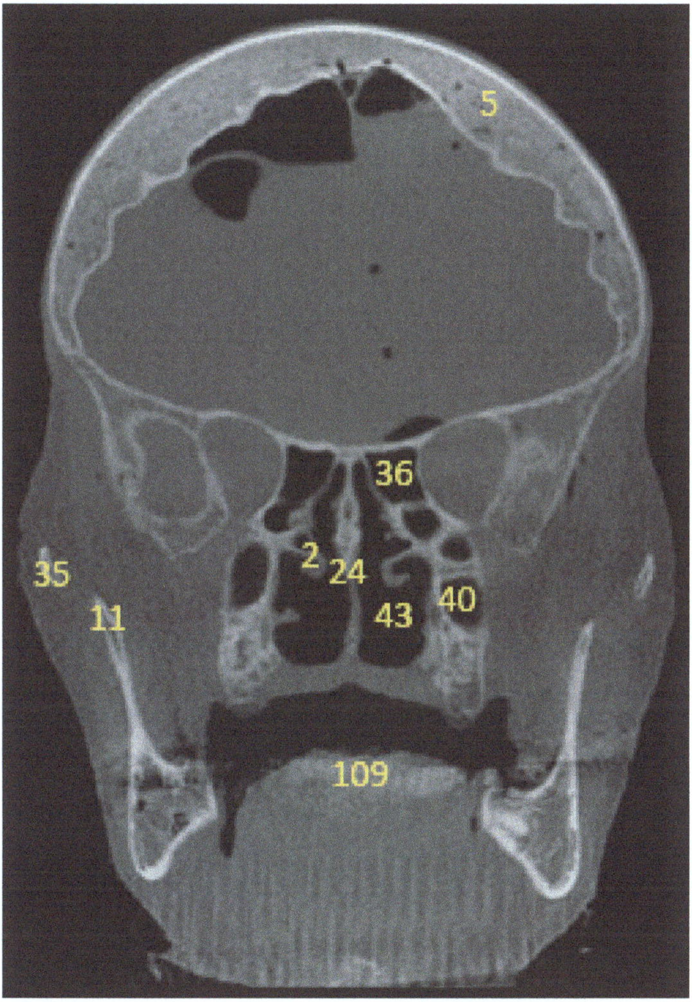

2 – Concha-middle	36 – Ethmoid sinus
5 – Frontal bone	40 – Maxillary sinus
11 – Mandible-coronoid process	43 – Nasal sinus
24 – Palatine bone-*vomer*	109 – Tongue
35 – Zygomatic arch	

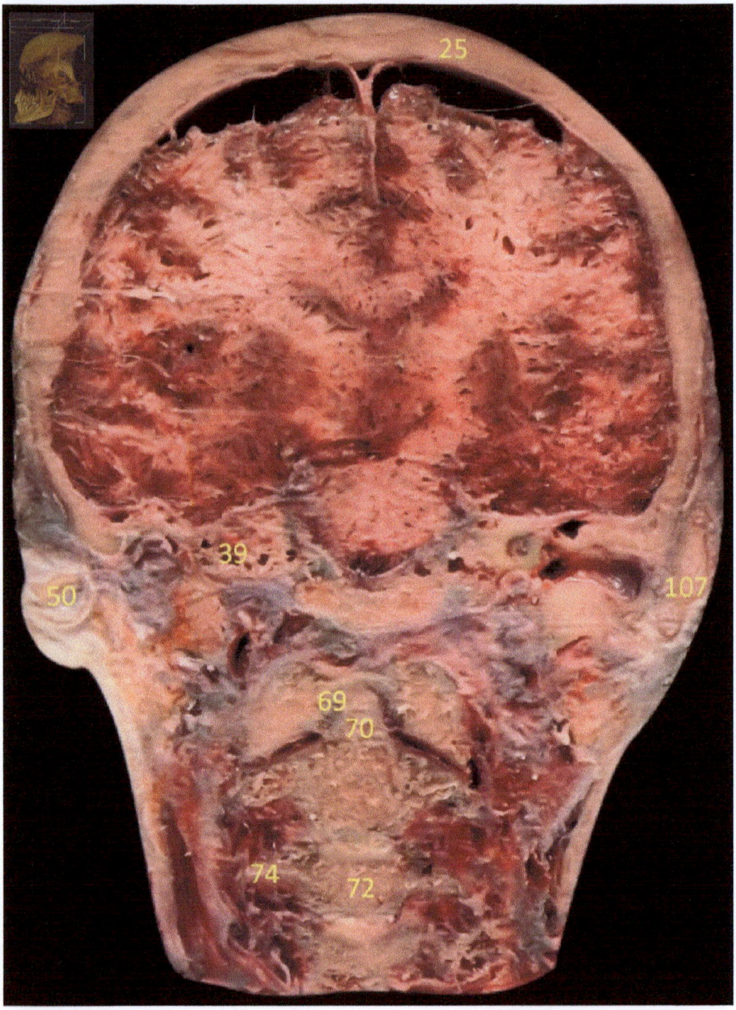

25 – Parietal bone	70 – C2 Axis-body
39 – Mastoid air cells	72 – C3-body
50 – Auricular canal	74 – C3-transverse process
69 – C2 Axis-dens	107 – Face

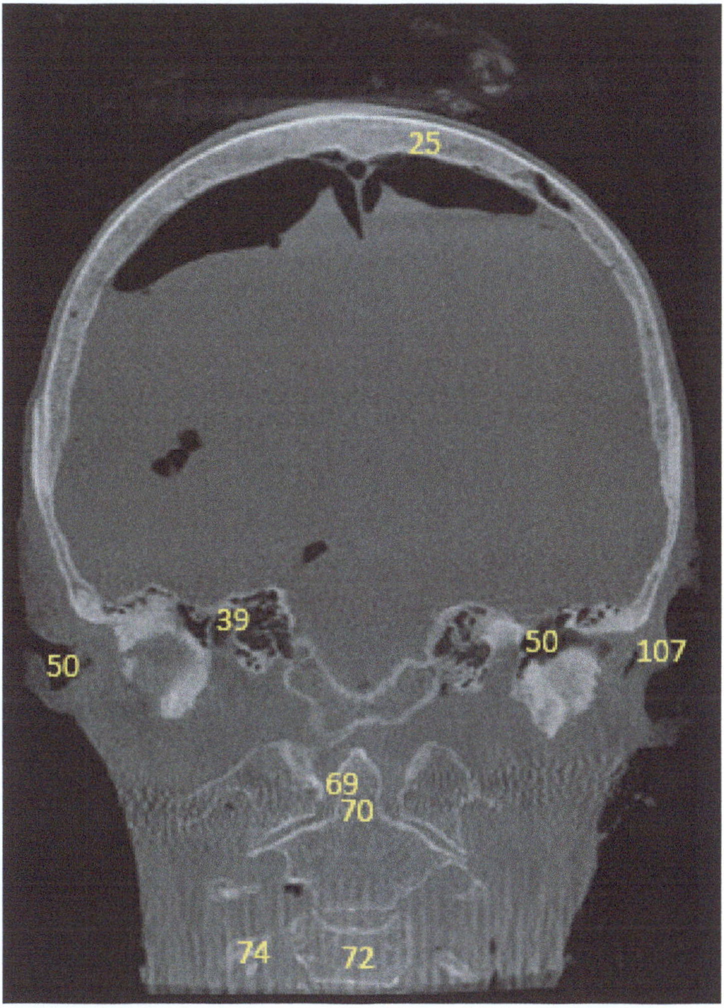

25 – Parietal bone
39 – Mastoid air cells
50 – Auricular canal
69 – C2 Axis-dens
70 – C2 Axis-body
72 – C3-body
74 – C3-transverse process
107 – Face

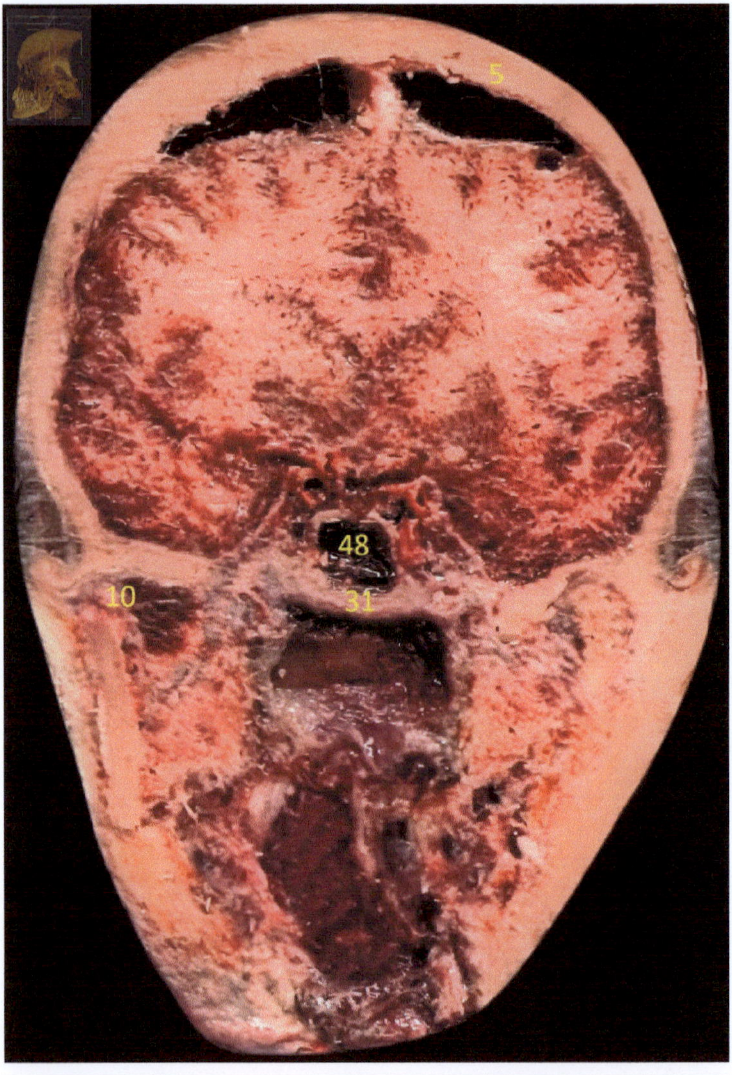

5 – Frontal bone
10 – Mandible-condyle
31 – Spehnoid bone
48 – Sphenoid sinus

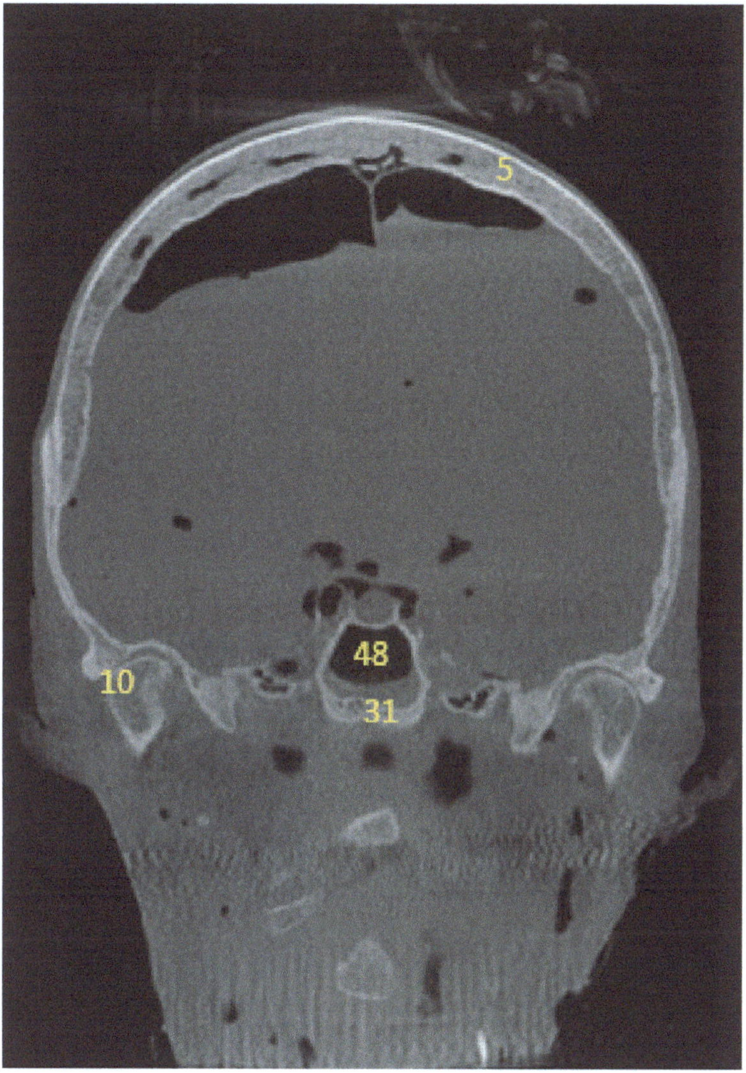

5 – Frontal bone
10 – Mandible-condyle
31 – Spehnoid bone
48 – Sphenoid sinus

Sagittal

C.H. Kau et al., *Cone Beam CT of the Head and Neck*,
DOI: 10.1007/978-3-642-12704-5_4, © Springer-Verlag Berlin Heidelberg 2011

Mid-Sagittal: Airspaces

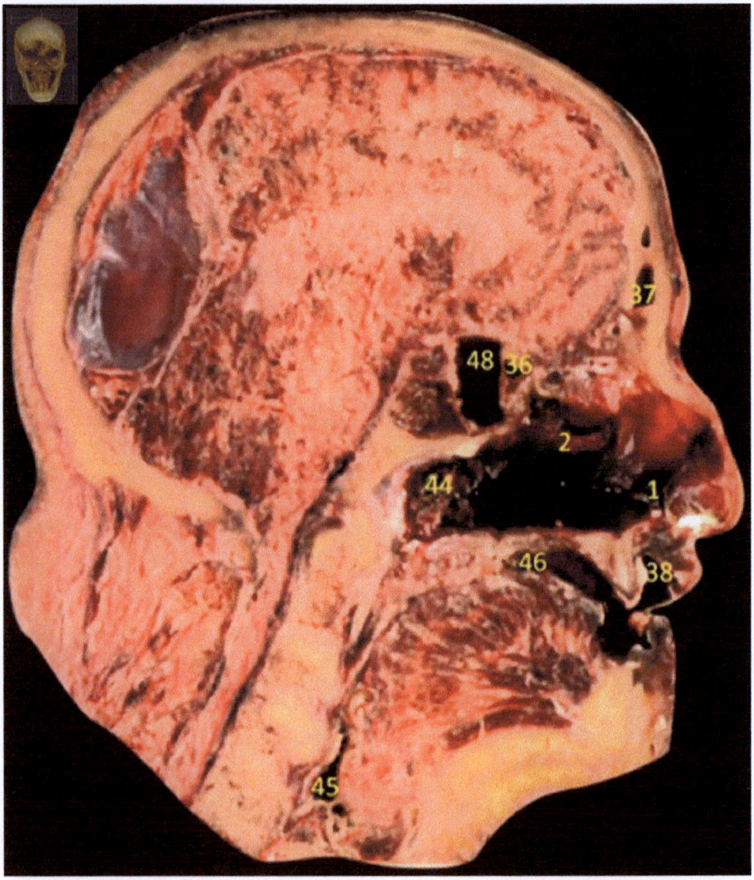

1 – Concha-inferior	44 – Nasopharyngeal airspace
2 – Concha-middle	45 – Oropharyngeal airspace
38 – Mandibular vestibule	46 – Palatoglossal airspace
36 – Ethmoid sinus	48 – Sphenoid sinus
37 – Frontal sinus	

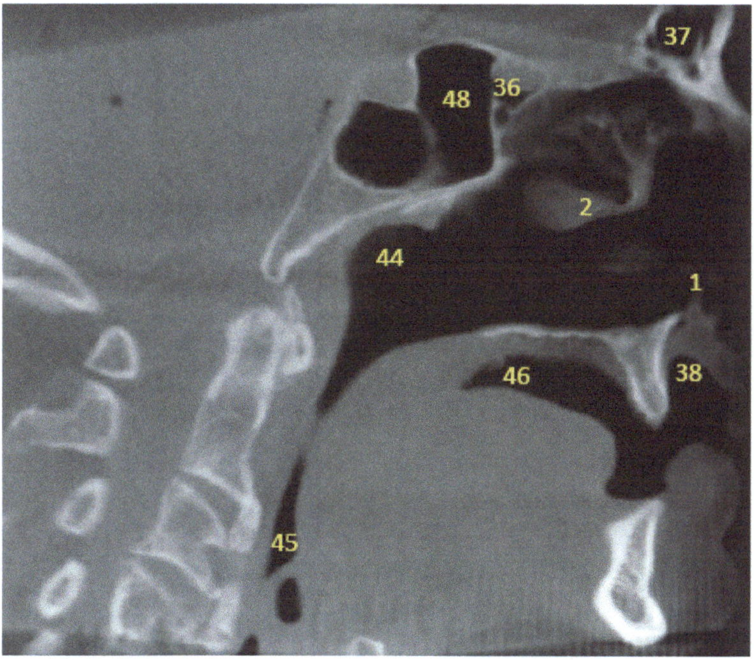

1 – Concha-inferior
2 – Concha-middle
38 – Mandibular vestibule
36 – Ethmoid sinus
37 – Frontal sinus
44 – Nasopharyngeal airspace
45 – Oropharyngeal airspace
46 – Palatoglossal airspace
48 – Sphenoid sinus

Mid-Sagittal Osteology

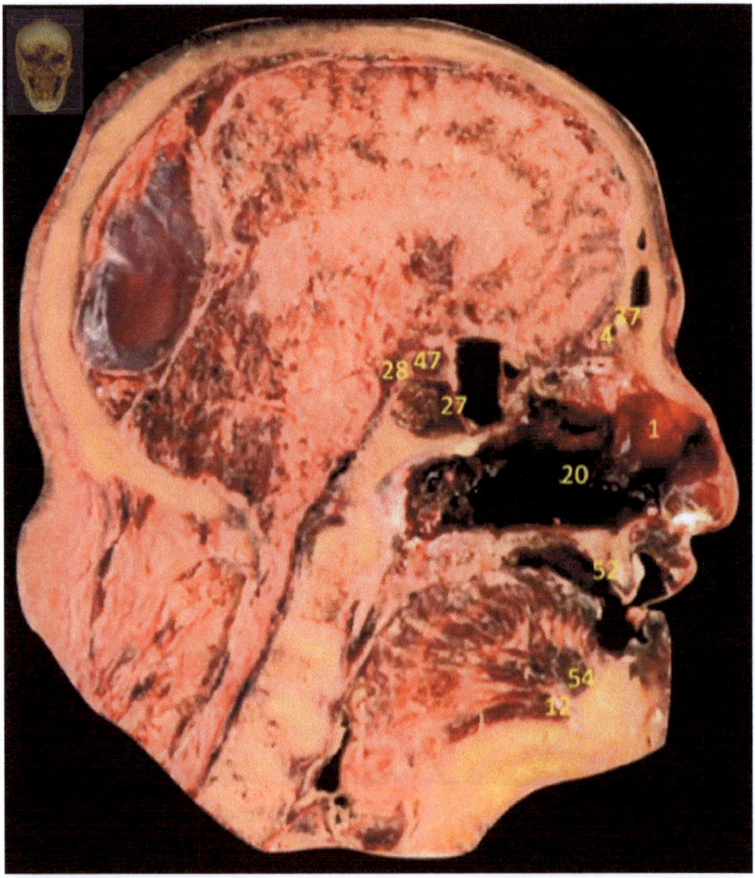

4 – Ethmoid bone-*crista galli*	37 – Frontal sinus
12 – Mandible-genial tubercle	47 – Pituitary fossa
20 – Nasal septum	52 – Incisive canal
27 – Septum in spehnoid sinus	54 – Lingual foramen
28 – Sphenoid-posterior clinoid process	

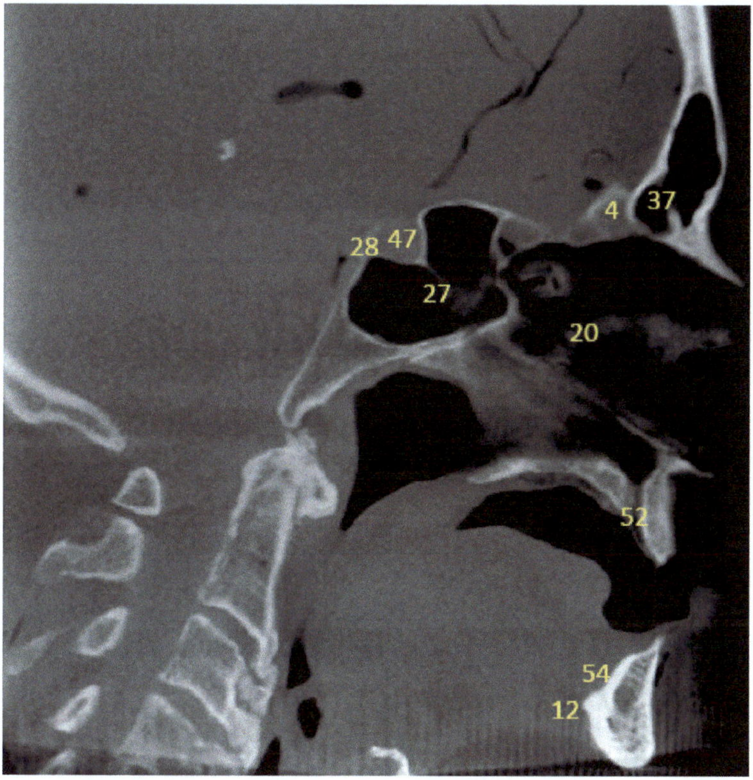

4 – Ethmoid bone-*crista galli*
12 – Mandible-genial tubercle
20 – Nasal septum
27 – Septum in spehnoid sinus
28 – Sphenoid-posterior clinoid process
37 – Frontal sinus
47 – Pituitary fossa
52 – Incisive canal
54 – Lingual foramen

2.5 cm from Mid-Sagittal

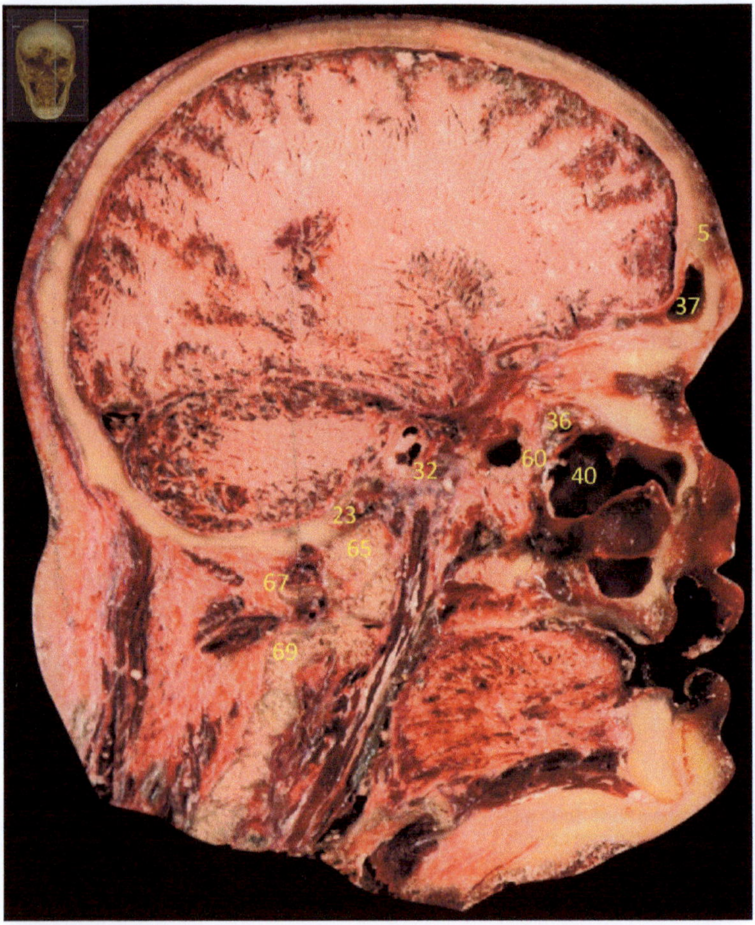

5 – Frontal bone	60 – Pterygomaxillary fissure
23 – Occiput-condylar process	65 – C1 Atlas-body
32 – Temporal bone	67 – C1 Atlas-posterior arch/tubercle
36 – Ethmoid sinus	69 – C2 Axis-dens
37 – Frontal sinus	
40 – Maxillary sinus	

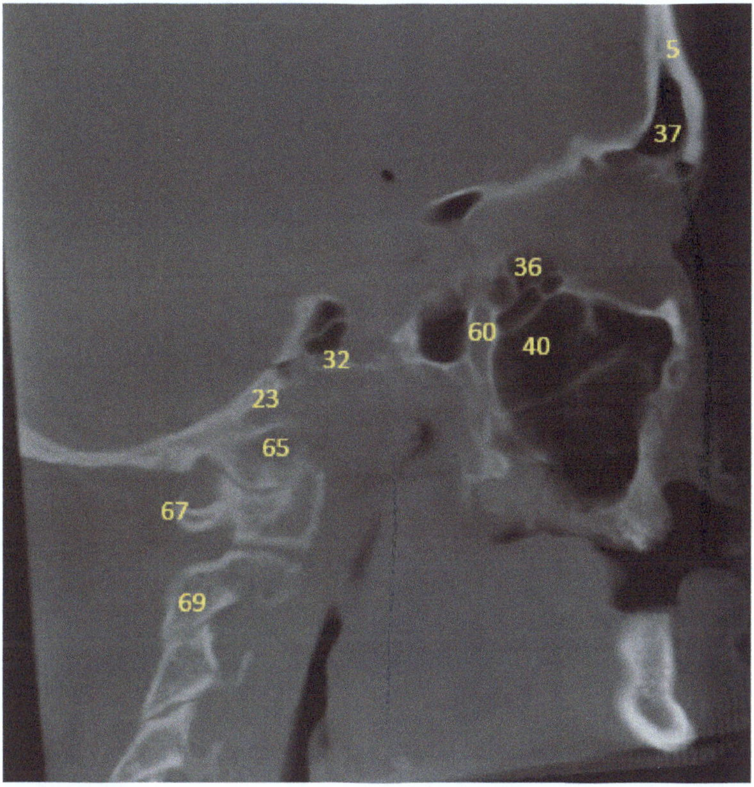

5 – Frontal bone
23 – Occiput-condylar process
32 – Temporal bone
36 – Ethmoid sinus
37 – Frontal sinus
40 – Maxillary sinus
60 – Pterygomaxillary fissure
65 – C1 Atlas-body
67 – C1 Atlas-posterior arch/tubercle
69 – C2 Axis-dens

5 cm from Mid-Sagittal (S2)

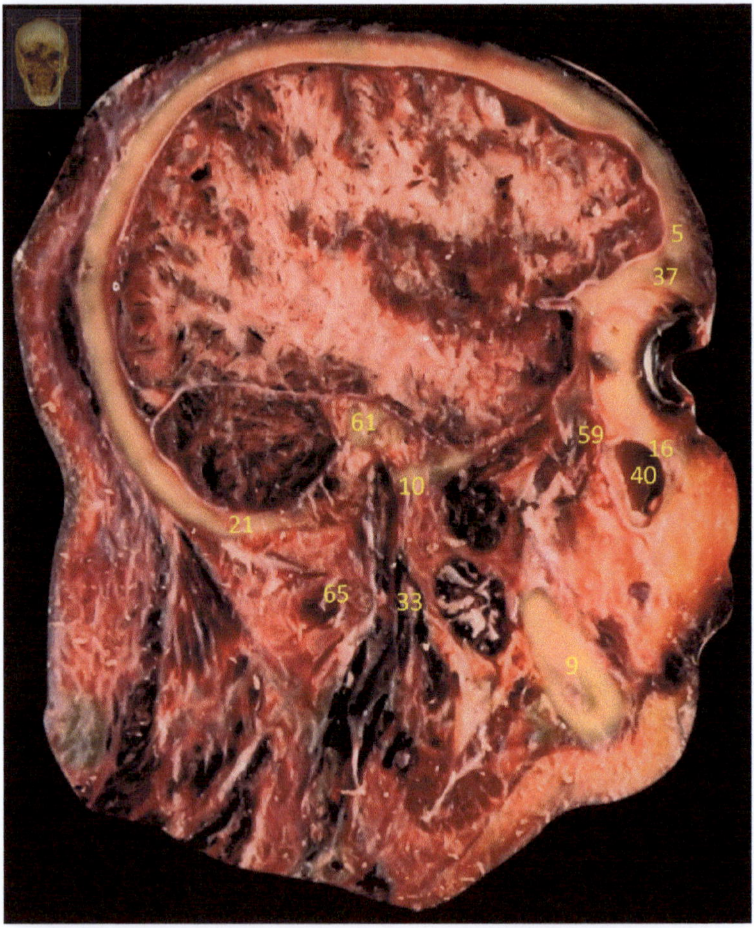

5 – Frontal bone
9 – Mandible-body
10 – Mandible-condyle
16 – Maxilla-zygomatic process
21 – Occiput
33 – Temporal bone-styloid process
37 – Frontal sinus
40 – Maxillary sinus
59 – Orbital fissure
61 – Temporal bone-petrous ridge
65 – C1 Atlas-body

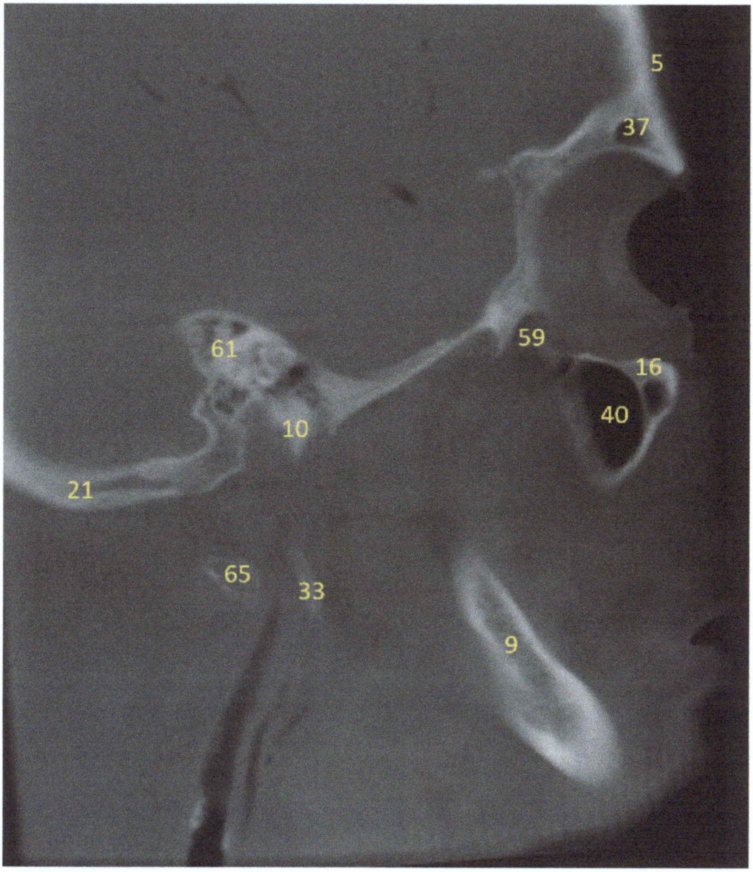

5 – Frontal bone
9 – Mandible-body
10 – Mandible-condyle
16 – Maxilla-zygomatic process
21 – Occiput
33 – Temporal bone-styloid process
37 – Frontal sinus
40 – Maxillary sinus
59 – Orbital fissure
61 – Temporal bone-petrous ridge
65 – C1 Atlas-body

7.5 cm from Mid-Sagittal

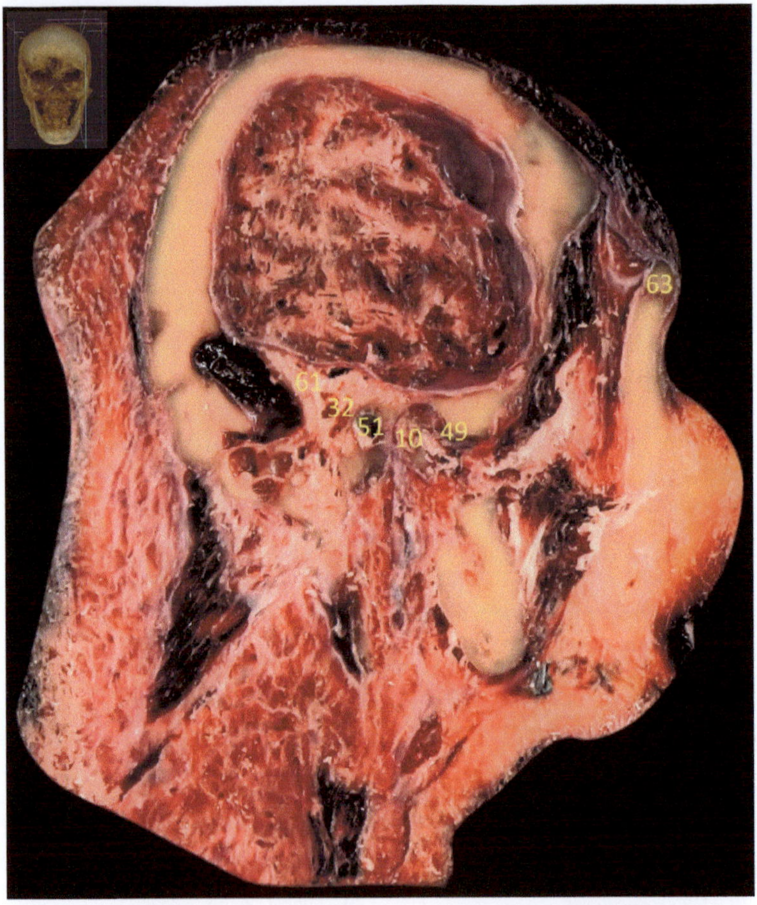

10 – Mandible-condyle
32 – Temporal bone
49 – Articular eminence
51 – External acoustic meatus
61 – Temporal bone-petrous ridge
63 – Zygomaticofrontal suture

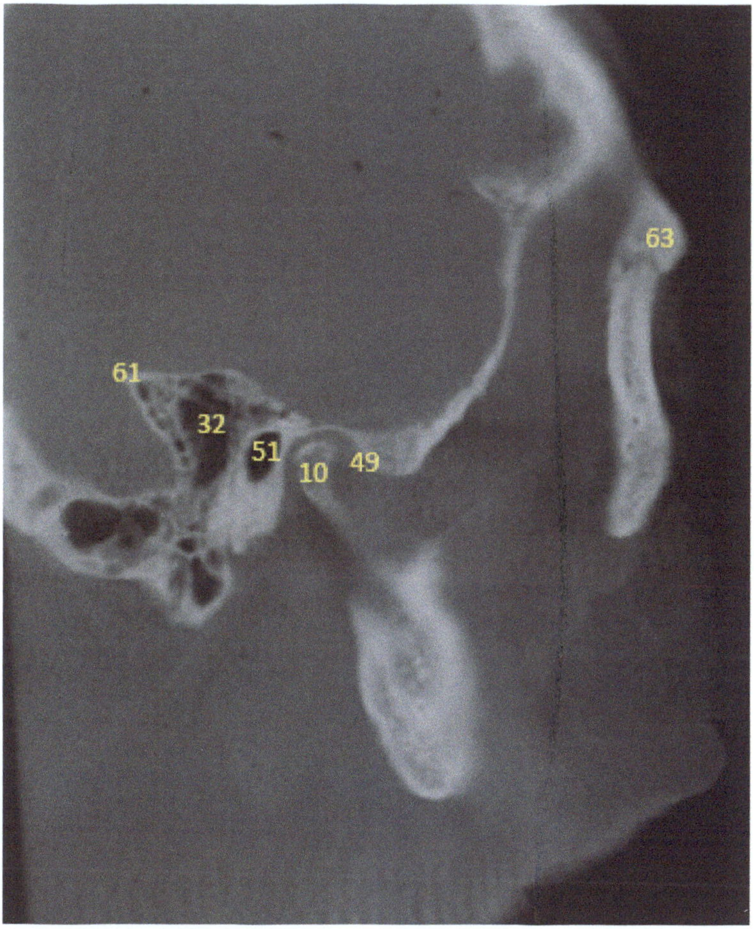

10 – Mandible-condyle
32 – Temporal bone
49 – Articular eminence
51 – External acoustic meatus
61 – Temporal bone-petrous ridge
63 – Zygomaticofrontal suture

10 cm from Mid-Sagittal

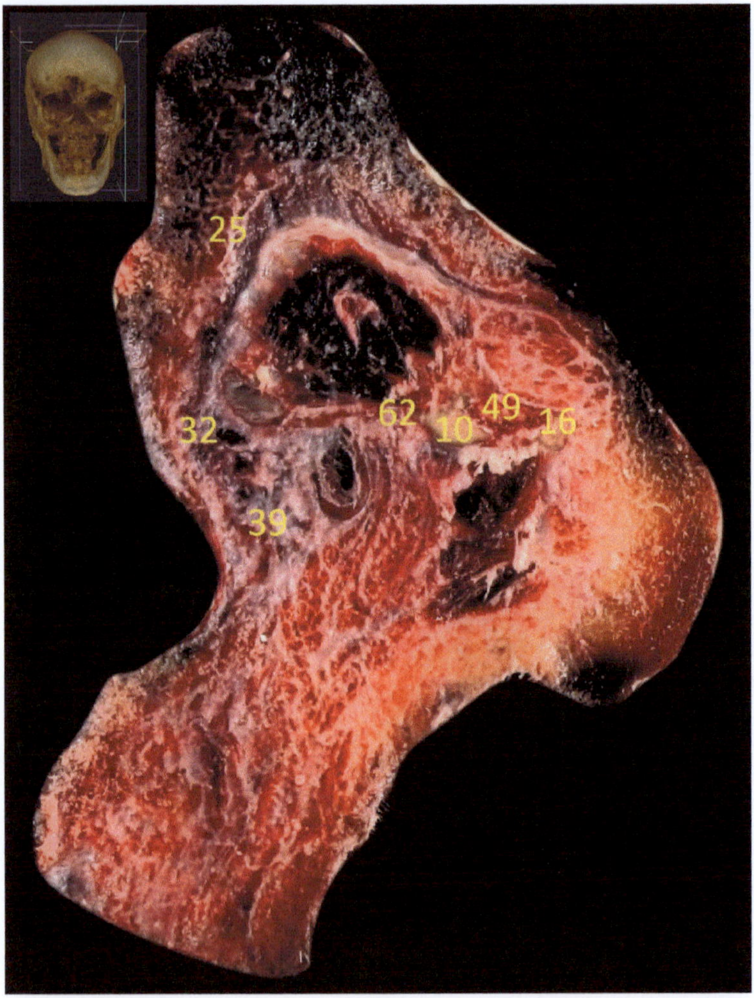

10 – Mandible-condyle	39 – Mastoid air cells
16 – Maxilla-zygomatic process	49 – Articular eminence
25 – Parietal bone	62 – Temporal fossa
32 – Temporal bone	

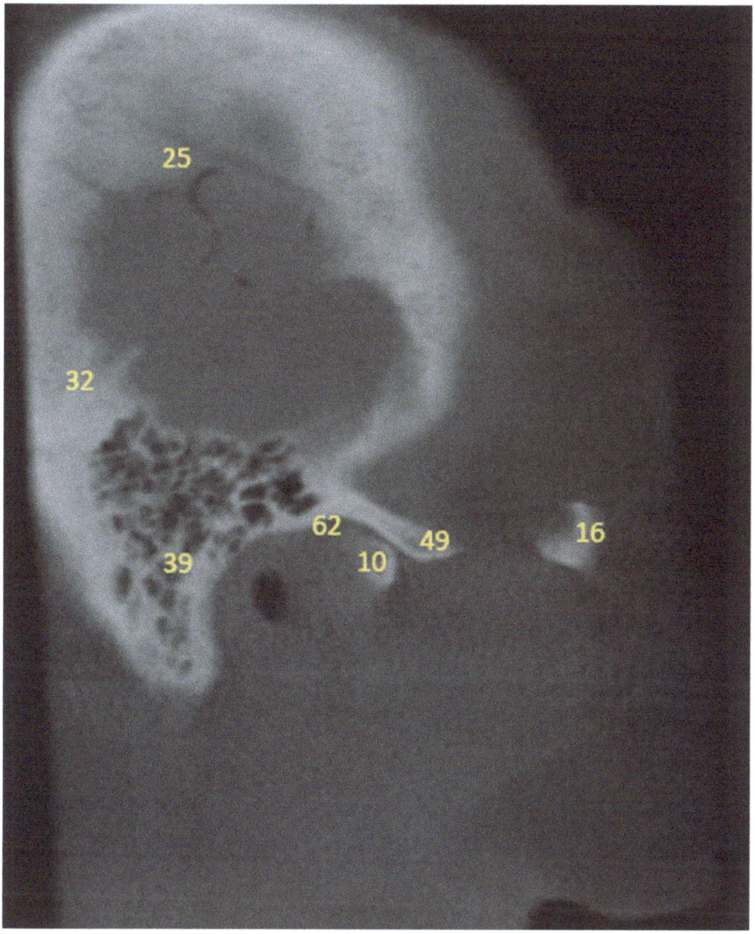

10 – Mandible-condyle
16 – Maxilla-zygomatic process
25 – Parietal bone
32 – Temporal bone
39 – Mastoid air cells
49 – Articular eminence
62 – Temporal fossa

34. Zygoma
35. Zygomatica rch
36. EthmoidSinus
37. FrontalSinus
38. Mandibularve stibule
39. Mastoid Air cells
40. MaxillarySinus
41. MaxillaryV estibule
42. Nasals eptum
43. NasalSinus
44. Nasopharyngeala irspace
45. Oropharyngeala irspace
46. Palatoglossala irspace
47. Pituitaryfos sa
48. SphenoidSinus
49. Articulare minence
50. AuricularC anal
51. External acoustic meatus
52. Incisivec anal
53. Internal Carotid artery
54. Lingualfora men
55. Mid palatal suture
56. Optic nerve canals
57. Oralc avity
58. Orbit
59. Orbitalfi ssure
60. Pterygomaxillaryfi ssure
61. Temporal Bone - petrous ridge
62. Temporalfos sa
63. Zygomaticofrontals uture
64. Zygomaticotemporals uture
65. C1 Atlas - body
66. C1 Atlas - anterior arch/tubercle
67. C1 Atlas - posterior arch/tubercle
68. C2 Axis - Transverse process
69. C2 Axis - Dens

70. C2 Axis - Body
71. C2 Axis - spine
72. C3 - Body
73. C3 - spine
74. C3 - transverse porcess
75. Mandibular Central Incisor Teeth (Crown)
76. Mandibular Central Incisor Teeth (Root)
77. Mandibular Cuspid (Crown)
78. Mandibular Cuspid (Root)
79. Mandibularfi rst bi-cuspid (Crown)
80. Mandibularfi rst bi-cuspid (Root)
81. Mandibularfi rst molar (Crown)
82. Mandibularfi rst molar (Root)
83. Mandibular Lateral Incisor Teeth (Crown)
84. Mandibular Lateral Incisor Teeth (Root)
85. Mandibular second bi-cuspid (Crown)
86. Mandibular second bi-cuspid (Root)
87. Mandibular second molar (Crown)
88. Mandibular second molar (Root)
89. Mandibular third molar (Crown)
90. Mandibular third molar (Root)
91. Maxillary Central Incisor Teeth (Crown)
92. Maxillary Central Incisor Teeth (Root)
93. Maxillary Cuspid (Crown)
94. Maxillary Cuspid (Root)
95. Maxillaryfi rst bi-cuspid (Crown)
96. Maxillaryfi rst bi-cuspid (Root)
97. Maxillaryfi rst molar (Crown)
98. Maxillaryfi rst molar (Root)
99. Maxillary Lateral Incisor Teeth (Crown)
100. axillary Lateral Incisor Teeth (Root)
101. Maxillary second bi-cuspid (Crown)
102. Maxillary second bi-cuspid (Root)
103. Maxillary second molar (Crown)
104. Maxillary second molar (Root)
105. Maxillary third molar (Crown)

106. Maxillary third molar (Root)
107. Face
108. Lips
109. Tongue

MIX
Papier aus verantwortungsvollen Quellen
Paper from responsible sources
FSC® C105338

If you have any concerns about our products,
you can contact us on
ProductSafety@springernature.com

In case Publisher is established outside the EU,
the EU authorized representative is:
Springer Nature Customer Service Center GmbH
Europaplatz 3, 69115 Heidelberg, Germany

Printed by Libri Plureos GmbH
in Hamburg, Germany